The LIBRARY of LANDFORMS™
ISLANDS

Isaac Nadeau

The Rosen Publishing Group's
PowerKids Press™
New York

Published in 2006 by The Rosen Publishing Group, Inc.
29 East 21st Street, New York, NY 10010

First Edition

Editor: Rachel O'Connor
Book Design: Elana Davidian

Photo Credits: Cover © Leonard Douglas Zell/Lonely Planet Images; p. 4 (top) © Yann Arthus-Bertrand/Corbis; p. 4 (inset) © 2002 GeoAtlas;
p. 4 (bottom) © Danny Lehman/Corbis; p. 4 (inset) © 2002 GeoAtlas; p. 7 (left) © Lee Foster/Lonely Planet Images; p. 7 (right) © Steve
Winter/National Geographic/Getty Images; p. 8 © Joe Skipper/Reuters/Corbis; p. 11 (top) © Bob Charlton/Lonely Planet Images; p. 11
(bottom) © Chris Mellor/Lonely Planet Images; p. 12 © Royalty-Free/Corbis; p. 15 (top) © Ralph Lee Hopkins/Lonely Planet Images; p. 15
(bottom) © Corbis; p. 16 © Douglas Peebles/Corbis; p. 19 (top) © Karl Lehmann/Lonely Planet Images; p. 19 (bottom left) © Woodward
Payne & Beverly Anderson/Lonely Planet Images; p. 19 (bottom right) © David Tipling/Lonely Planet Images; p. 20 © Kennan Ward/Corbis.

Library of Congress Cataloging-in-Publication Data

Nadeau, Isaac.
  Islands / Isaac Nadeau.
    p. cm. — (Library of landforms)
  Includes index.
  ISBN 1-4042-3126-9 (lib. bdg.)
  1. Islands—Juvenile literature. I. Title.

GB471.N34 2006
551.42—dc22

                    2005001557

Manufactured in the United States of America

CONTENTS

Atlantic Ocean   Ireland   Irish Sea

Top: The country of Ireland is an island that measures 32,636 square miles (84,527 sq km). The Irish Sea is on its eastern coast and the Atlantic Ocean is on its west. Shown here is part of the island's western coastline. *Bottom:* Some islands are so small they do not even have a name, such as this one near Tortola in the British Virgin Islands.

4

What Is an Island?

Do you know what Greenland, Hawaii, and New York's Manhattan have in common? The answer is they are all islands. Islands come in all shapes and sizes. There are so many islands on Earth that no one has been able to count all of them! An island is a body of land that is surrounded on all sides by water. Islands are found in lakes, rivers, seas, and oceans. Continents are the largest landmasses on Earth, and they are surrounded by water. They are not considered islands because they are so large. The only difference between an island and a continent is size. Australia is the smallest continent, measuring about 2.9 million square miles (7.5 million sq km). Greenland is the world's largest island. Greenland is about 840,000 square miles (2.2 million sq km) in area. Not all islands are so large. All over the world there are thousands of tiny, unnamed islands. The smallest islands measure less than 1 square foot (.09 sq m).

The state of Hawaii consists of a chain of islands. The largest island in the chain is Hawaii, or the Big Island, as it is also known. Hawaii is the largest island in the United States. It measures about 4,021 square miles (10,414 sq km).

How Islands Are Formed

Islands are formed in many different ways. Some islands are formed by **volcanoes**, such as the Hawaiian Islands and the Aleutian Islands in Alaska. Other islands, known as barrier islands, are found near sandy beaches. The movement of wind and waves piles the sand offshore, making islands. The Outer Banks of North Carolina is a chain of barrier islands. Another type of island is a coral reef island, which is made up mostly of coral. The Florida Keys, off the southern coast of Florida, is an example of a chain of coral reef islands. Every island is different, but islands can be separated into two main groups. These are continental islands and oceanic islands. Oceanic islands are islands that rise to the surface from the ocean floor. These islands are usually volcanic. Continental islands are formed when a body of water rises and separates a piece of land from the continent. Greenland is an example of a continental island. Greenland is connected underwater to the North American continent by a **ridge** that is only 600 feet (183 m) below the surface.

Left: The Outer Banks in North Carolina is a chain of barrier islands. The Outer Banks is an example of a continental island chain. *Right:* The Florida Keys is a chain of coral reef islands. The chain forms an archipelago, or group of islands. Islamorada, shown here, is one of the main islands of the Upper Keys.

4.5 BILLION YEARS AGO:	4.4–3.8 BILLION YEARS AGO:	3.5–3.8 BILLION YEARS AGO:	ABOUT 600 MILLION YEARS AGO:	1.8 MILLION TO 11,000 YEARS AGO:	TODAY:
Earth forms. It is so hot that there is only water vapor, not liquid. Therefore there are no oceans, rivers, or lakes. No islands exist yet. Molten rock cools, forming the first solid rock.	Earth cools enough for liquid water to form on the surface. The first islands were probably formed at this time.	Life on Earth begins.	The first coral reefs begin to be formed in the oceans. As ocean levels rise and fall throughout the years, some of these coral reefs become islands.	Earth goes through an ice age. Glaciers advance and retreat, so sea levels fall and rise. This causes many islands throughout the world to grow and shrink.	Volcanoes continue to erupt in the Pacific Ocean and elsewhere, forming new islands or adding to older ones. Coral reefs continue to grow. All over the world, islands change in small and large ways every day.

Pictured here is the Cape Hatteras Lighthouse, found on one of the islands in the Outer Banks in North Carolina. In 1999, the lighthouse had to be moved because the shoreline around it had eroded so badly. Since the lighthouse was built in 1870, more than 1,000 feet (305 m) of beach have eroded away.

Barrier Islands

Barrier islands are a type of continental island. Barrier islands are the result of **erosion** and **deposition**. Erosion occurs when wind and waves along the shoreline break down shells, cliffs, and rocks into small bits of **sediment** and carry them away. Deposition occurs when this sediment is dropped to the ground by the wind or sinks to the ocean floor. Barrier islands are made of sand and sediment that are deposited by wind and ocean waves. Barrier islands are usually long and narrow and run along the coastline.

Barrier islands can be found on the eastern coast of North America from New Jersey to Mexico. Most of these barrier islands were created during a rise in sea level that occurred about 18,000 years ago. The sea level rose because a rise in **temperature** caused ice from **glaciers** to melt and pour into the oceans. As the sea level rose, many of the beaches along the coastlines were flooded. The ocean waves dragged sand from these beaches out to sea, forming large piles of sand. Some of the piles of sand rose above the water, forming barrier islands.

Coral reef islands are usually found in warm, **shallow** ocean water with a lot of sunlight. The Florida Keys, the Bahamas, and many islands in the Caribbean Sea are examples of coral reef islands. Coral reefs are large, stony shapes that are formed by tiny animals called polyps. A polyp is a type of animal that feeds on bits of food that float in the ocean water. They build the coral reefs as a home under water. The reefs are formed as polyps lay down small amounts of **minerals** every day. Each year these reefs grow higher and higher. A drop in sea level can leave parts of the reefs above water, forming islands. A drop in sea level can occur when temperatures on Earth cool and water freezes, forming large glaciers. Large coral reefs can take millions of years to form. When a reef is above water, other plants and animals may come to live on it over time. As plants and animals grow and die on these new islands, their bodies rot and mix with minerals in the coral. This forms a **layer** of soil. The soil allows many different kinds of plants to grow.

Top: The Great Barrier Reef, off the northeast coast of Australia, is the largest coral reef in the world. It is 1,250 miles (2,012 km) long. Pictured here is Lady Musgrave Island in the Great Barrier Reef. *Bottom:* Under water, coral reefs are full of life. In this picture you can see soft coral polyps in the center of the yellow coral.

The island of Manhattan in New York City is surrounded by the waters of the Hudson River, the East River, the Harlem River, and Upper New York Bay. Here you can see Manhattan's East River. The rocks that make up Manhattan are similar to the rocks on the mainlands of New York and New Jersey nearby. These rocks include very old rocks such as granite, as well as newly deposited sedimentary rocks left by glaciers about 20,000 years ago.

RIVER AND LAKE ISLANDS

Islands found in rivers and lakes are continental islands. They share the same rock types as the continents to which they are connected. By looking at the rocks of an island, **geologists** can figure out whether it was formed in the same way and at the same time as the mainland. For example, by studying the rocks of Orleans Island in eastern Canada, geologists have learned that the southern part of the island is formed from the same rocks as the Appalachian mountain chain. These mountains run along the East Coast from Canada to Alabama. They were formed hundreds of millions of years ago and were once much higher than they are today. Over millions of years of erosion, the Appalachian Mountains have become smaller and smaller. About 20,000 years ago, a huge sheet of ice covered most of Canada and much of the United States. When the ice sheet melted, many lakes and rivers were formed, including the Saint Lawrence River. The Saint Lawrence River began to flow around the ancient mountains of the Appalachians, surrounding some of the old, worn-down **peaks** with water, including Orleans Island.

Volcanic Islands

Many islands are formed by volcanic activity. Volcanoes occur when the **molten** rock beneath Earth's surface is forced upward through a weak spot in Earth's **crust**. The crust of Earth is made up of several large plates that fit together like pieces of a puzzle. Some of these plates are lighter in weight than others. The lighter plates are called continental plates. Continental plates include almost all Earth's land. The heavier plates are oceanic plates, which make up the ocean floors.

When volcanic **eruptions** occur on oceanic plates, they can result in the formation of islands. Sometimes these eruptions must occur many times, building up layer on layer of new rock before they reach above the surface of the ocean. Mount Haleakala, a volcano in Hawaii, helped form the island of Maui. Mount Haleakala rises more than 30,000 feet (9,144 m) from the bottom of the ocean. Its peak is 10,032 feet (3,058 m) above the surface of the sea. Most volcanic islands are found in the Pacific Ocean, where there is a lot of volcanic activity.

Top: The Galapagos Islands are volcanic islands found in the Pacific Ocean. Here you can see Isla Bartolome. At .6 square miles (1.5 sq km), it is one of the smaller islands of the Galapagos.
Bottom: In this overhead view of the Galapagos Islands you can see Isla Isabela, which is the largest of the Galapagos Islands. Running the length of the picture, Isla Isabela measures 1,774 square miles (4,595 sq km).

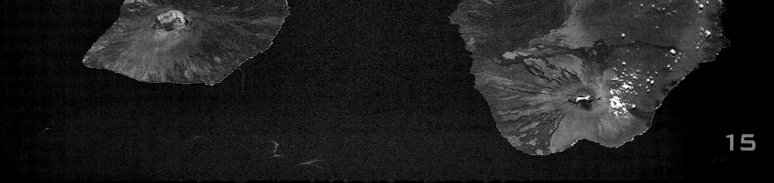

Reaching 13,794 feet (4,204 m) above sea level on the Big Island, Mauna Kea is the tallest volcano in Hawaii. *Mauna Kea* is a Hawaiian name meaning "white mountain." The top of Mauna Kea is cold enough for snow to fall on it in the winter, making it appear white. Mauna Kea has not erupted for about 4,500 years, but many scientists believe that it will erupt again one day.

THE HAWAIIAN ISLANDS

The islands of Hawaii are probably the most famous and most-studied islands in the world. They are also the farthest from any continent. The Hawaiian Islands are an archipelago. An archipelago is a chain of islands that are close together. The Hawaiian Islands are volcanic. They were formed by volcanoes that have been erupting from the ocean floor for several million years. There are about 130 islands that make up the state of Hawaii. They were all formed by underwater volcanoes. Most of these islands are small, but there are eight major islands. The biggest island is called Hawaii, which is also known as the Big Island. The other islands include Maui, Kahoolawe, Lanai, Molokai, Oahu, Kauai, and Niihau. Not all the islands are the same age. The Hawaiian Islands have been formed over about the last 70 million years. The Big Island is the youngest, at about 1 million years old. Two of the volcanoes on the Big Island, Kilauea and Mauna Loa, are active today. Every time these volcanoes erupt, new land is added to the islands.

LIFE ON ISLANDS

Scientists have studied life on islands for many years. They have learned about a few basic trends in life on islands. In general islands that are young usually have fewer **species** than do older islands. Old islands have more species because plants and animals have had more time to reach them. When a volcanic island first rises from the ocean, there is no life on it. Over time birds may fly across the water and land on the island. Also the wind may carry seeds of plants to the island. Scientists have also found that large islands usually have more life on them than small islands have. This is because there are more places for plants and animals to live. In addition, scientists have found that islands close to the mainland usually have more species than do faraway islands. The island of Madagascar is close to the African continent and has thousands of species of plants and animals. Whereas, the island of Saba in the Caribbean, which is far from any mainland, has **relatively** few species.

Although the Hawaiian Islands are far from any mainland, they have existed for about 70 million years and are rather large. This allows them to be home to many species of plants and animals. By some counts there are more than 20,000 species of living things in the Hawaiian Islands and the waters around them. Of these species almost 9,000 are found nowhere else on Earth.

Top: The black and white lemur can only be found on the island of Madagascar. Because these lemurs are in danger of dying out, they are protected by the government. *Bottom Left:* One of the many beautiful flowers found on the Big Island of Hawaii is the colorful bird of paradise. *Bottom Right:* A lot of horned puffins make their homes on islands in the North Pacific. The Pribilof Islands, off the coast of Alaska, are home to the horned puffin pictured here.

Midway Island is a volcanic island in the Pacific Ocean. It was formed nearly 70 million years ago. However, there is no longer any volcanic activity on the island. This means that no new land is being added to the island, and erosion can take place without interruption. Here a bird flies over the island, which is now nearly at the same level as the sea.

How Islands Change

Islands do not stay the same from day to day. Erosion and deposition play a part in changing the landscape of islands. For example, on sandy barrier islands, ocean waves erode the islands' beaches. Another example of erosion occurs on the cliffs of volcanic islands. Waves crash against the cliffs, breaking off pieces of rock and carrying them away. Some volcanic islands, such as Midway Island in the Pacific Ocean, have eroded so that they are nearly at sea level. In perhaps a few million years, Midway Island may disappear underwater.

Volcanoes on some islands continue to erupt, adding new rock as the lava cools. For example, between 1983 and 2000, about 506 acres (205 ha) of land were added to the Big Island as a result of the eruptions of Kilauea. In other places new islands are being formed. Just offshore of the Big Island in Hawaii, there is an active volcano called Loihi. Loihi is 3,178 feet (969 m) below sea level. Some scientists expect that, with more eruptions, Loihi will break the surface and become an island within the next hundred thousand years.

People and Islands

Throughout history islands have held a special place in people's hearts. In oceans and lakes all over the world, people have set out by boat to explore islands. In many cases when people discovered islands, they stayed to live there. In fact 700 million people live on about 9,000 islands. This means that one in every nine people lives on an island.

Many people visit islands on vacation. On islands in the Caribbean and on many other islands around the world, **tourism** is an important business. People enjoy islands because of their natural beauty, their beaches, and their interesting plant and animal life. People also visit islands to study plants, animals, volcanoes, and rocks. However, scientists have learned that many living things on islands are in danger of becoming extinct, or dying out. This is because their **habitats** are being destroyed when people or other species move in. To protect native species, it is important to protect their surroundings. The more people learn about islands, the better we are at understanding how we can take care of these treasured places surrounded by water.

GLOSSARY

crust (KRUST) The outer, or top, layer of a planet.

deposition (deh-puh-ZIH-shun) The dropping of sediment by wind or water.

erosion (ih-ROH-zhun) The wearing away of land over time.

eruptions (ih-RUP-shunz) Explosions of gases, smoke, or lava from volcanoes.

geologists (jee-AH-luh-jists) Scientists who study the form of Earth.

glaciers (GLAY-shurz) Large masses of ice that move down a mountain or along a valley.

habitats (HA-bih-tats) The surroundings where an animal or a plant naturally lives.

layer (LAY-er) One thickness of something.

minerals (MIN-rulz) Natural elements that are not animals, plants, or other living things.

molten (MOL-ten) Made liquid by heat.

peaks (PEEKS) The very tops of things.

relatively (REH-luh-tiv-lee) As compared to something else.

ridge (RIJ) The long, narrow, upper part of something.

sediment (SEH-deh-ment) Gravel, sand, or mud carried by wind or water.

shallow (SHA-loh) Not deep.

species (SPEE-sheez) A single kind of living thing.

temperature (TEM-pruh-cher) How hot or cold something is.

tourism (TUR-ih-zem) A business that deals with people who travel for pleasure.

volcanoes (vol-KAY-nohz) Openings in Earth's surface that can shoot up molten rocks, called lava.

INDEX

WEB SITES

Due to the changing nature of Internet links, PowerKids Press has developed an online list of Web sites related to the subject of this book. This site is updated regularly. Please use this link to access the list:
www.powerkidslinks.com/liblan/islands/